The King Tiger Tank

by

Horst Scheibert

Schiffer Military History
Atglen, PA

Sources

Federal Archives
Koblenz Podzun-Pallas
Publishers Archives
Scheibert Archives
Profile Publications
Egon Kleine P. F. Strauss

Translated from the German by Dr. Edward Force.

Library of Congress Catalog Number: 89-084175.

Printed in the United States of America.
ISBN: 0-88740-185-6

Published by Schiffer Publishing Ltd.
77 Lower Valley Road
Atglen, PA 19310
Please write for a free catalog.
This book may be purchased from the publisher.
Please include $2.95 postage.
Try your bookstore first.

A maintenance crewman cleans the big road wheels with a spray gun. This Tiger II also has the later type of turret.

A Tiger II, also called "King Tiger", with the turret used on the first fifty tanks. It has a skirting plate, and the turret is bowed out on the left side for the commander's seat.

Armored Battle Tank VI (Tiger II), Type B

As ordered by Hitler, the Henschel firm and Professor Porsche were commissioned in the spring of 1941 to develop a battle tank with a fighting weight of about forty tons and correspondingly heavy armament. When the German troops encountered the Russian KW I and II and the T 34 that same year, the level of urgency was raised.

Thus in the spring of 1942 two prototypes were presented: the technically more conservative tank made by the Henschel firm proved to be the better one; the Porsche prototype, on the other hand, which was equipped with many technical innovations, failed completely. Nevertheless Hitler, probably because of the reputation of the renowned tank constructor, ordered both types developed further.

Thus the Tiger I of Henschel and the "Ferdinand" (later renamed "Elephant") of Professor Ferdinand Porsche came into being. Both were equipped with 8.8 cm tank cannon (Kwk).

Since both tanks were unable to stay at the required weight, the Tiger I weighed 60 tons, the Ferdinand 70, it was decided to build a new heavy tank. Thus the Panzer V, called "Panther", came about, weighing approximately 40 tons.

As early as the spring of 1943, after the Tank IV (Tiger I) had seen its first action and the "Panther" had proved ready for production, not to mention the planning of a "Panther II", the Weapons Office requested a thoroughgoing equalization of construction groups to simplify the preparation process and the spare-parts situation. Since the Tiger I's first action showed that more shot-deflecting forms had to be added, a new Tiger type was developed, the Tiger II, also called the "King Tiger" (first by the enemy, then by the Germans too). It went into series production at the Henschel works at the beginning of 1944. The inclusion of a longer 8.8 cm cannon (L/71 instead of the L/56 of the Tiger I) required different running gear. All this, in addition to somewhat stronger armor, increased the weight by eleven tons over that of the Tiger I. But with this weight of about 70 tons, not only was the vehicle underpowered (with an unfavorable power-to-weight ratio), but it also exceeded its usefulness. There were only a few bridges that could carry its weight, and special vehicles and chains had to be built to transport it.

Its main external difference from the Tiger I is its better-looking turret. There were two versions of it. The first fifty used the so-called Porsche turret, which was originally intended for the Porsche Tiger. From the 51st tank on, the King Tiger was fitted with a new turret (called the production turret) made by the firm of Krupp. This turret stood out because of its strengthened front armor and smaller front surface. In addition, it allowed the carrying of 84 shells, six more than before. It can also be distinguished from the Porsche turret by the lack of a bowed-out left side for the commander's cupola.

The Tiger II was utilized in independent groups (heavy armored units), which were put under the temporary command of armies or corps as strong-point weapons. It saw action almost exclusively in the West.

There were also "King Tiger" command cars. But they differed only in having more radio equipment, at the cost of the aforementioned quantity of ammunition. In addition, a "Pursuit Tiger", developed from it, was built (some of which were armed with a 12.8 cm antitank gun). It had a non-turning turret.

The Porsche turret of the "King Tiger", production units 1-50.

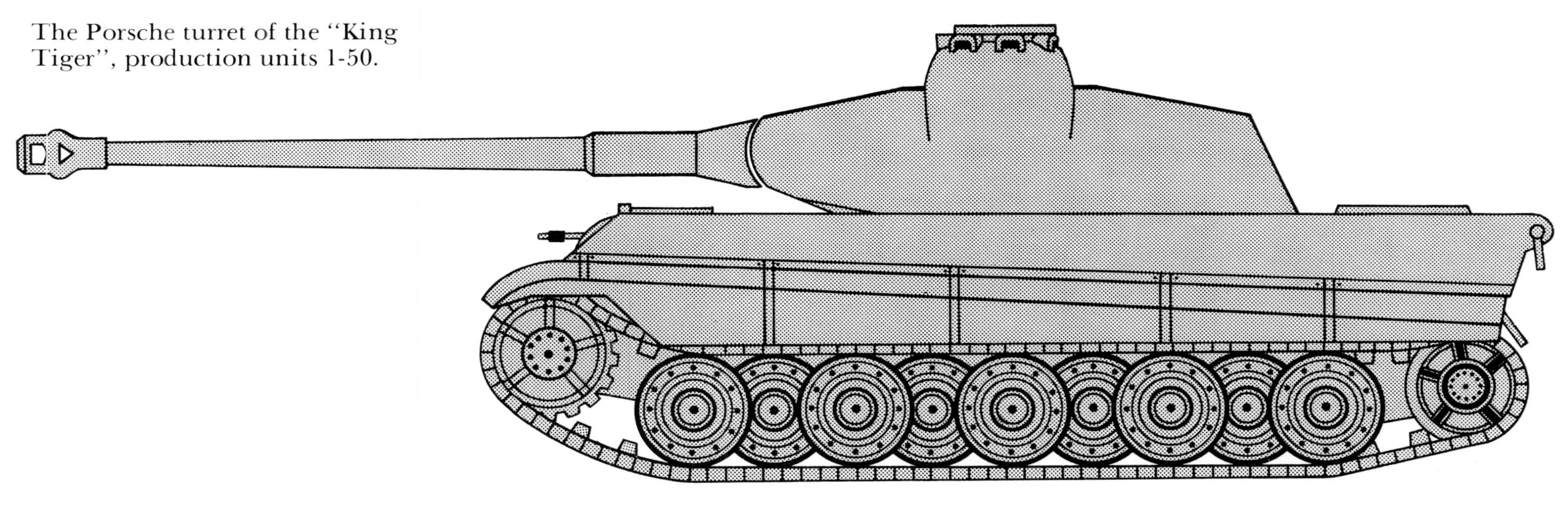

The so-called "Production Turret" (developed by Krupp) of the further "King Tigers".

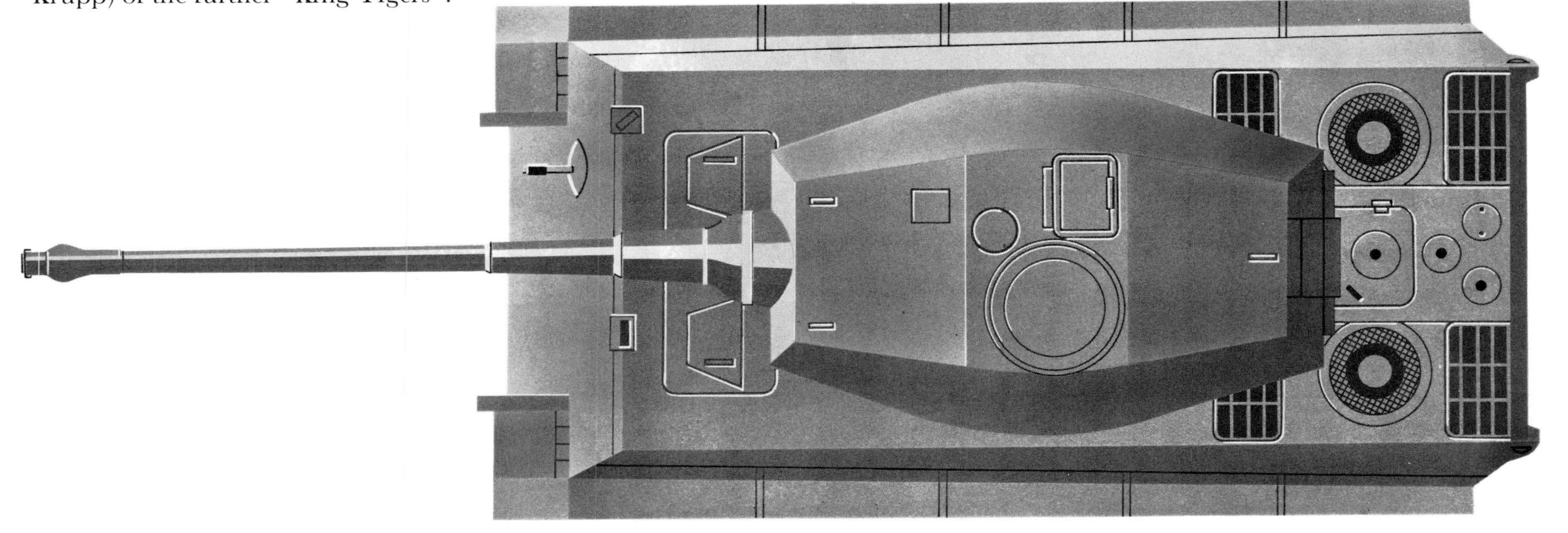

The "King Tiger" attained legendary fame, somewhat unjustifiably. On the one hand, for reasons already mentioned; on the other, because of the too-long, "unhandy" cannon, which gave its shells a higher penetrating power but caused greater wear and tear and, because of its instability, had poorer accuracy than the L/56. It also had to be adjusted very frequently. The interior space of the Tiger II was very limited, more so than the Tiger I or Panzer V. But there is no doubt that the "King Tiger", despite all its teething problems that quick developments always show, and the "Pursuit Tiger" developed from it all the more, were superior to all other tanks in the world at that time. Its chief disadvantages were its too-high weight and, above all, its too-small numbers.

In all, there were built:

Sd.Kfz 182	"Tiger II" Type B "Tiger II"	Hen- schel	1944-1945 487
Sd.Kfz 267/268	Armored Command Car "Tiger II" Type B	Hen- schel	1944-1945
Sd.Kfz 186	Pursuit Tank "Pursuit Tiger" Type B	Nibe- lun- gen werke	1944-1945 70
	Armored Recovery Vehicle "Tiger II"	Alkett rebuilt	1944-1945 18

Views of the Tiger II with "Production Turret"

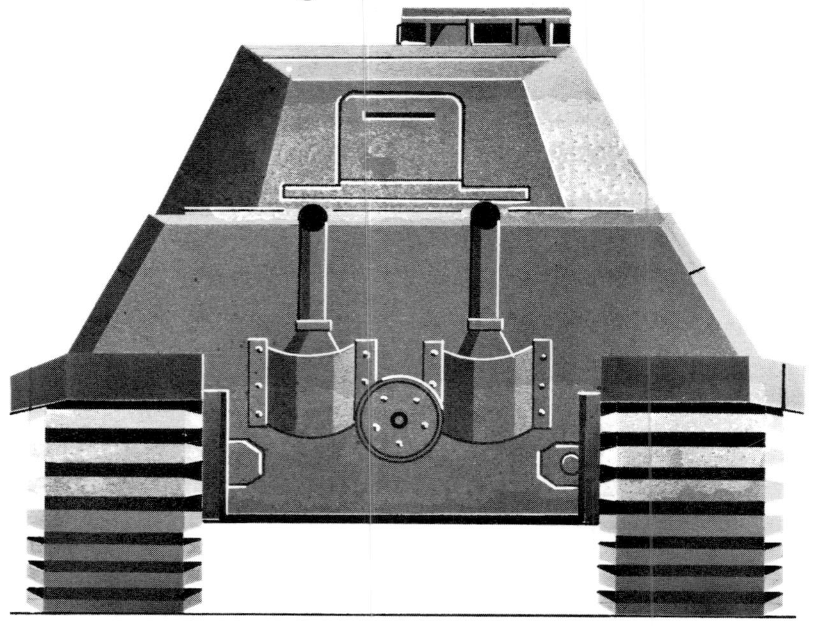

Technical data of the motor and transmission of the Tiger II

Motor	As of Unit 251, Maybach HL 230 P 45 carburetor engine
Cylinders	12 (60-degree V-form)
Bore & stroke	130 x 145 mm
Displacement	23,880 cm.
Maximum power	700 HP at 3000 rpm
Steady power	600 HP at 2500 rpm
Compression	4 Solex 52 JFF II D double downdraft cross-country carburetors
Valves	Dropped. 1 camshaft per cylinder head, driven by spur gear
Main bearings	8. Replaceable cylinder liners
Cooling	water, via pump
Batteries	2 12-volt 150 ah
Generator	1000 W
Starter	6 HP
Transmission	rear engine, driving chain tracks, semi-automatic pre-selector gearbox
Gearbox	Maybach Olvar 40 12 16, 8 forward speeds, 4 reverse speeds
Ratio	Lateral gearing 10.7

Further technical data at the end of this volume.

The driver's seat.

Racks for the 8.8 cm ammunition shells, at right behind the radioman.

Tiger II with "Porsche turret" at the Bergen firing range in the Lüneburg Heath, 1944.

Tiger II with the rolled turret. Its cross emblem is not painted on the turret, but on the hull. The 8.8 cm cannon (L/71) is equipped with a muzzle guard.

5 Tiger II tanks, type B, with "Porsche turrets", at firing practice on the Bergen firing range. The front tank has lost part of its track skirting.

The same Tiger II unit seen from the other side. The power loading of the shells for the longer 8.8 cm cannon (Kwk) was stronger than that of the 8.8 cm Kwk of the Tiger I (L/56). Unfortunately, there was no possibility of exchanging ammunition.

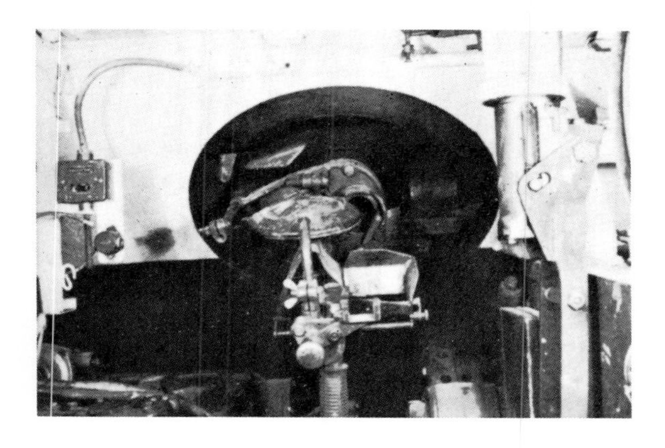

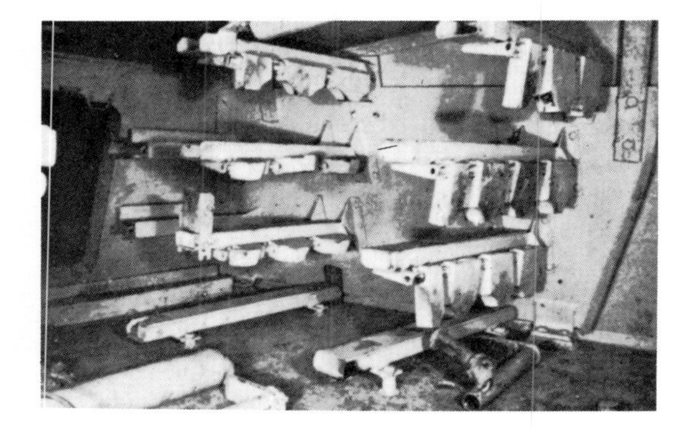

Upper left: Close-up view of the commander's cupola with its seven viewing ports. Recognizable are (at left) the mounting for the anti-aircraft gun and the turret ventilator (foreground).
Above: The length of the cannon and the bowed-out form of the commander's seat of the "Porsche turret" are easy to recognize. The bow machine gun is extended.
Center: The inside of the bow machine gun bullet shield can be seen from the bow gunner's seat.
Lower left: The picture shows the shell racks in the left rear of the turret.

The Tiger II could destroy all armored vehicles of its time at 2000 meters, but was itself impervious at this distance thanks to its frontal steel armor of 150 and 185 mm thickness.

This photo shows the rippled Zimmerite covering of the steel plates. It was intended to prevent the adhesion ofmagnetic explosives. The turret painting consisted of three colors: a sand-colored background with greenish and brown areas.

The leading tank has just fired a shot. The heavy smoke and the more or less heavy dust cloud, depending on the ground surface, prevented target-hitting observation.

Here is a Tiger II with "Porsche turret" and a different camouflage pattern. Its spots of color are smaller than those on the tanks previously shown. Commander and officer observe the terrain.

Both photos show the same heavy Tiger unit (equipped with Tiger II "Porsche turrets") concealed behind the invasion front. At right, captured English soldiers help distribute supplies.

The Tiger II was heavier than the Tiger I, over a meter longer, 23 cm higher, but only 2 cm wider.

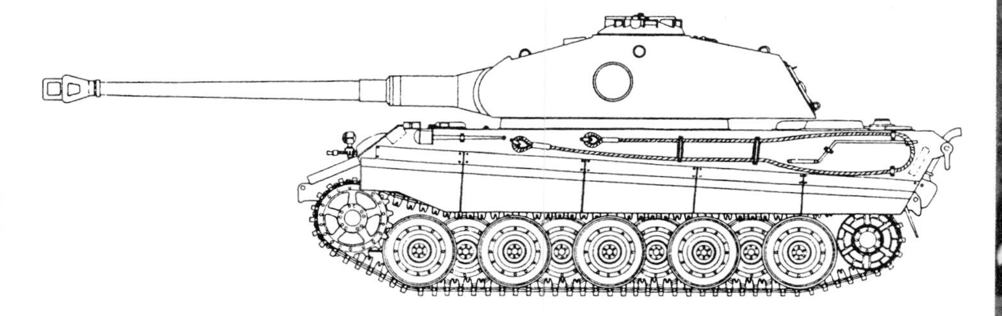

The shape of the Tiger II was similar to that of the "Panther" and must be regarded as particularly successful.

An abandoned "King Tiger" in Normandy, 1944. It is examined by English soldiers above, by Americans below. The Tiger II was usually lost on account of technical failure, less often from shot damage. In the retreat fighting, which included almost all fighting from 1943 on, technical failure often meant the loss of the vehicle. Depending on the urgency of the situation, the ordered explosion was often impossible. The King Tiger shown here seems to have been lost because of technical failure.

A good picture of a Tiger II with the new turret. This so-called "Production turret" was fitted to all Tiger II units produced from the 51st on. It was built by Krupp. The Zimmerite overlay, some of which has already been lost, is readily recognized here.

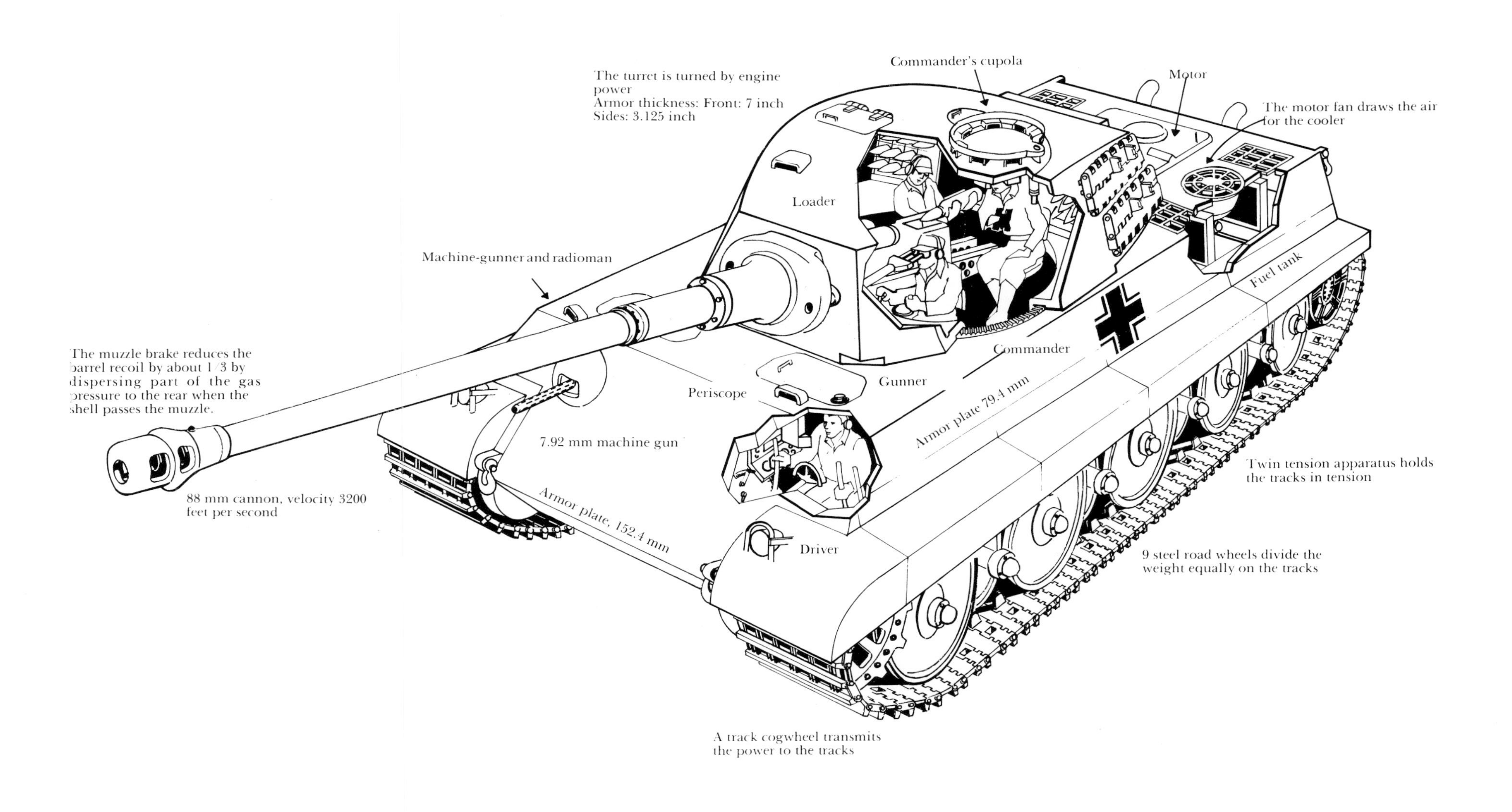

Diagonal sketch of a King Tiger with the later turret, to show the positions of the crew.

The "King Tiger" also used spare track links as extra armor. Beneath the cannon the bow machine-gunner (also radioman) looks out of his entrance hatch. He could also raise his seat; this indicates that this Tiger II is not in battle.

On the next 11 pages are excerpts from weekly newsreels. Though some of them have technical defects, this is because of the very great enlargement and the age of the films. We considered these photos, though, to be very important, as they show many details of equipment. They are from Heavy Tiger Unit 503 which, after being supplied with "King Tigers", demonstrated formations and firing drills at the Sennelager firing range. The tanks have differing camouflage paint patterns and, as yet, no numbers.

Technical service of a "King Tiger". This photo shows that the newer turret with its "pig-head visors" (others of which are seen on the assault guns in Volume 4) and conical form, which offered the enemy an even smaller target than the "Porsche turret" of the first 50 Tiger II tanks produced.

Here are two more photos from the weekly newsreels. The unit has driven up and awaits the command of the officer seen in the foreground.

Lifting the right arm means "attention!"; it is followed by the sign of the action to be taken, for example: "March!", "Hold back!", etc. Naturally, these commands could also be given by radio. During radio silence (for example, on the march into battle) and (as here) for display purposes, hand signals were used.

Officers of a Heavy Armored Unit (503). They can be recognized by the silver piping on their black caps and their shoulder insignia.

Authentic action photos of "King Tigers" are rare. Here are two from a Tiger II unit in southern Poland (autumn of 1944).

The Tiger II had two different types of camouflage painting. Both were three-colored: tan, green and brown. But within these colors, some had particular spots, either light on a dark background or dark on light. Here a "King Tiger" has its paint work touched up with a spray gun.

A Tiger II with its wide fighting tracks. Its width measured 80 cm, as opposed to the railroad shipping tracks of only 60 cm. With the former, the ground pressure (weight per square centimeter) amounted to 1.06 kg.

A "King Tiger" meets captured Americans on its way to the front. Battle of the Ardennes, December 1944.

A Tiger II with paratroopers riding on it during the Battle of the Ardennes, December 1944-January 1945.

This Tiger II shows the somewhat narrower shipping tracks. They were sometimes also used in battle, and the later camouflage paint with additional spots of color on the flat surfaces.

A Tiger II with scars after a battle. The "wound" inside the circle is a typical glancing blow, thanks to the slanted armored surface.

The brackets on the turret and hull served to hold spare track links, towing lines and hand tools such as shovels, wire cutters, wrenches, axes etc.

A Tiger II with the narrower shipping tracks.

This abandoned Tiger shows the rear hatch of the turret, in this case still a Porsche turret. Ammunition was loaded through it, and if necessary (for repairs), the cannon could be taken out through it.

For the "King Tiger" there was a special low loader, which was used to take abandoned Tiger II tanks back to the workshops. They were designed so that they could take a tank even with fighting tracks.

This and the following photos show the "King Tiger" that stands on the grounds of the Battle Troop School II at the Munster base in the Lüneburg Heath. In this photo too, the rear turret hatch, here on a "production turret", is easy to see.

The exhaust system of the "King Tiger", like those of most tanks, is pointed upward, so as to allow as great a fording depth as possible.

Here it can be seen clearly how far the fighting track extends over the width of the road wheels. This achieved a favorable ground pressure, to be sure, and thus allowed better travel on soft ground, but made "throwing a track" easier in heavy country.

This photo shows the sleek, almost elegant form of the "King Tiger". The hooks on the turret held spare track links.

The muzzle brake prevented the too-strong recoil of the barrel when firing. But in the narrow turrets of the tanks, unlike those of anti-aircraft guns, there was no room for them.

The profile of the heavy track links was particularly successful, not only in maintaining ground pressure but also in gaining adhesion on a slippery surface.

The brackets on the side of the hull were used to hold towing cables and hand tools, such as crowbars, axes, shovels etc.

Technical Data

Chassis and Body	Self-supporting armored hull, armored body with turning turret powered by vehicle's motor
Running Gear	Two chain tracks of 96 links each (130 mm intervals). Drive wheel forward, steered wheel aft. 9 large road wheels in a line.
Steering and Brakes	Hydraulic two-wheel steering controlled by steering wheel. Drive wheels with hydraulic Argus disc brakes

General Data

Track	2790 mm, with shipping tracks 2610 mm
Track width	800 mm, shipping tracks 600 mm
Overall dimensions	7260 x 3625 x 3090 mm. With cannon and skirting plates: 10286 x 3755 x 3090 mm. Overall width for shipping 3270 mm.
Armor	Front 100 x 185 mm, sides and rear 80 mm
Ground clearance	485 mm
Fording capability	1600 mm
Turning circle	5 meters diameter
Allowable gross wt.	68,000 kg
Top speed	40 kph
Fuel consumption	Road: 680 liters per 100 km, cross-country 1000 liters per 100 km.
Fuel capacity	860 liters (7 tanks)
Range	Road: 120 km, cross-country: 80 km.
Crew	5 men
Armament	8.8 cm Kwk 43 L/71 and 2 machine guns

Here the weighty mass of the turret can be seen in comparison to the hull. It resulted in, among other things, the high weight of about 70 tons and, as a result thereof, the unfavorable power-to-weight ratio (HP per ton) of 10.1 to 1. The present-day "Leopard" has a power-to-weight ratio of 20.8 to 1.

As fine as its form was, its motor was too weak for its weight, its fuel consumption was markedly greater than the Tiger I, and its cannon was too long and thus too sensitive, as well as being a hindrance in town and woodland fighting. The Hunting Tiger (right page) lacked a turning turret and usually had a more powerful cannon (12.8 cm Pak 44 L/55).

Hunting Tiger

Technical Data

Chassis and Body	Self-supporting armored hull, armored body with turning turret powered by vehicle's motor.
Bore & stroke	130 x 145 mm
Displacement	23, 880 cc
Maximum power	700 HP at 3000 rpm
Steady power	600 HP at 2500 rpm
Compression ratio	
Carburetors	2 Solex 52 JFF II D double downdraft country
Valves	Dropped, 1 camshaft per bank, gear driven
Main bearings	8, replaceable cylinder liners
Cooling	Water-cooled, with pump
Batteries	2 12 V 150 Ah
Generator	1000 W
Starter	6 HP
Power Transmission	Rear engine. chain-drive, semi-automatic pre-selector gearbox
Gearing	Maybach Olvar 40 12 16 gearbox, 8 forward, 4 reverse gears
Transmission	Spur gearing 10.7
Chassis and Body	Self-supporting armored hull, armored body without turning turret
Running gear	2 tracks, each of 96 links (130 mm intervals) front drive wheel, rear steered wheel, 9 large road wheels in straight line 2 suspension rods per pair of road wheels

Steering and Braking	Hydraulic two-wheel steering controlled by steering wheel; steered wheels have Argus hydraulic disc brakes

General Data

Track length	4240 mm
Track	2790 mm, 2610 mm with shipping tracks
Track width	800 mm, shipping tracks 600 mm
Overall dimensions	7800 x 3625 x 2945 mm; with cannon: 10654 x 5625 x 2945 mm
Armor	Front 200 mm, sides & rear 80 mm
Ground clearance	480 mm
Fording capability	1750 mm
Turning circle	5 meter diameter
Allowable gross wt.	75,000 kg
Top speed	35 kph
Fuel consumption	Road 700, cross-country 1000 liters per 100 km
Fuel capacity	860 liters
Range	Road 120, cross-country 80 km
Crew	6 men
Armament	12.8 cm 44 L/55 AA gun, (at times also 8.8 cm 43/2 L/71 AA gun) plus 1 machine gun

The Twelfth S.S. Armored Division Herbert Walther. This highly effective division is presented through hundreds of photographs and maps to record their personnel, vehicles, and operations, especially their activity in the Normandy invasion in 1944 and 1945.
Size: 7 3/4" x 10 1/2" 3 color maps 120 pp.
33 color photographs 233 b/w photographs
ISBN: 0-88740-166-X hard cover $24.95

The First S.S. Armored Division Herbert Walther. A pictorial history of the people, vehicles and activities of this important elite division which was active between 1939 and 1945.
Size: 7 3/4" x 10 1/2" 3 maps 120 pp.
339 b/w photographs 11 drawings
ISBN: 0-88740-165-1 hard cover $24.95

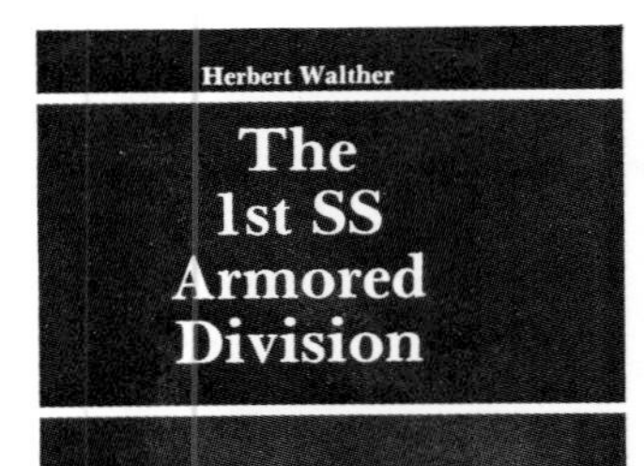

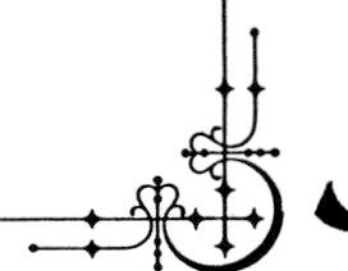

Schiffer Military

The Infantry Regiments of Frederick the Great, 1756-1763 Günter Dorn and Joachim Engelmann. The most famous field regiments and garrison regiments of Frederick's army in a splendid, large volume. Impressive color drawings and informative text. Regimental chronicles, lists of regimental commanders, etc.
Size: 9" x 12" 160 pp.
73 color illustrations
ISBN: 0-88740-163-5 hard cover $95.00

The Cavalry Regiments of Frederick the Great, 1756-1763 by Günter Dorn and Joachim Engelmann. Large format, color volume about Frederick's cavalry regiments. All cuirassier, dragoon and hussar regiments. Two color-illustrated pages for each regiment and informative text including the history of their formation to their dissolution.
Size: 9" x 12" 160 pp.
70 color illustrations 1 map
ISBN: 0-88740-164-3 hard cover $95.00

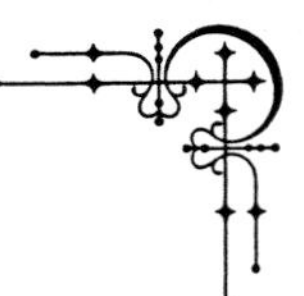

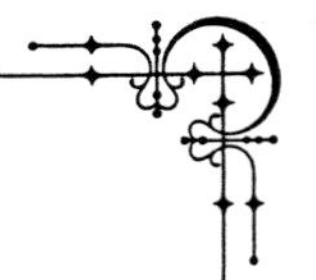

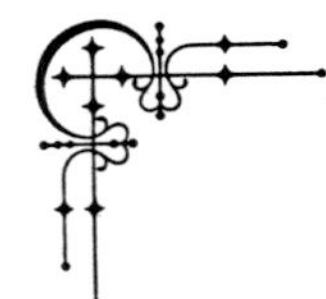

The Volkswagen ""Kübelwagen"in the War Reinhard Frank. A pictorial survey of the light, medium and heavy personnel vehicles of the German Army which are described and illustrated.
Size: 12" x 9" 52 pp.
Over 100 illustrations
ISBN: 0-88740-162-7 soft cover $9.95

The Leopard Family of West German Armored Vehicles Michael Scheibert. This fascinating book contains many illustrations of the Leopard I and II tanks and their derivatives such as tank recovery vehicles, armored bridging vehicles, engineering tanks and anti-aircraft tanks.
Size: 12" x 9" 8 diagrams 52 pp.
4 color illustrations 116 b/w photographs
ISBN: 0-88740-167-8 soft cover $9.95